edZOOcatio

Spiders Ev

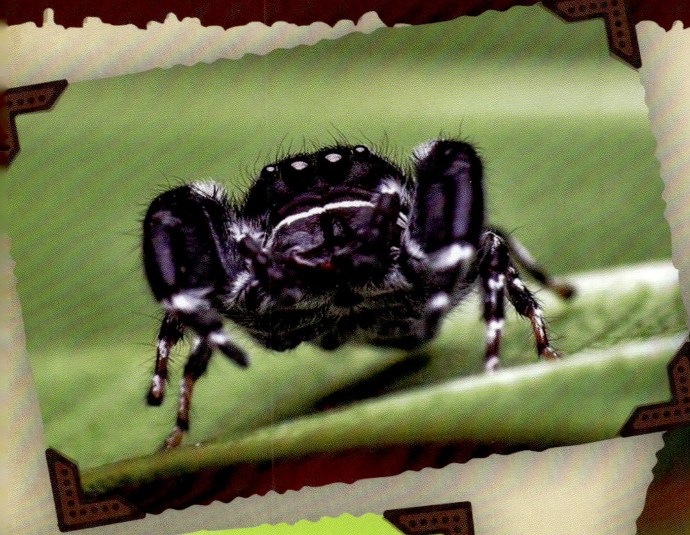

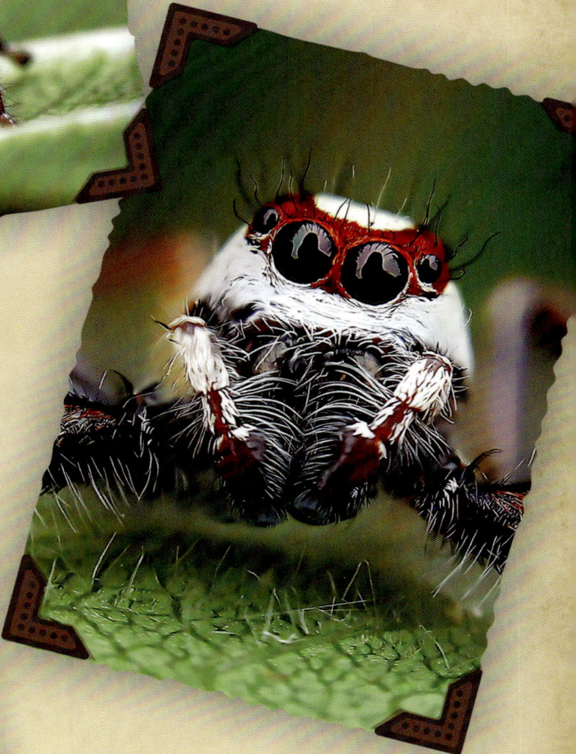

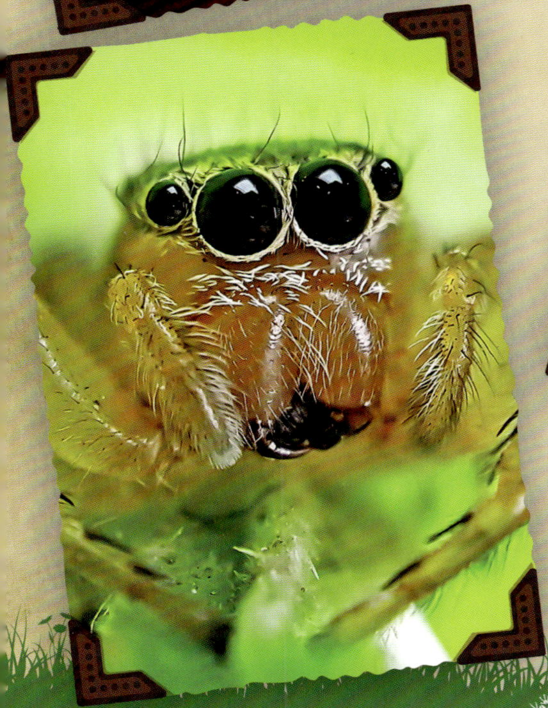

ZOOKEEPER
AGES 3-5

Wildlife Tree
edZOOcation

Spiders are tiny creatures with eight legs.

They can be found almost everywhere!

There are many kinds of spiders.

Some are big, and some are small.

Spiders have eight legs.

Count them with me!

Some spiders spin webs to catch their food.

Webs are sticky and strong.

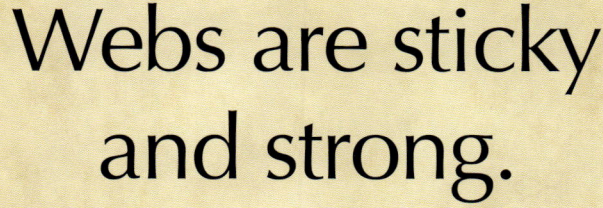

Spiders make different shapes of webs.

Some are round, and some are like sheets.

Spiders have many eyes.

Some can see very well!

Baby spiders are called spiderlings.

They hatch from eggs.

Some spiders are venomous, but most are harmless.

Always watch but don't touch!

Spiders help us by eating pests.

They keep our gardens healthy.

Spiders come in many colors.

Some are red, blue, or even green!

Spiders can be round, long, or even flat.

They have different shapes.

Goodbye, Spiders!

Dedication:

To the Xerces Society, for protecting invertebrates and their homes.

Thank you for your dedication.

- Jenny Curtis

For Jennifer

—A.R.

Curtis, Jenny. Assisted by OpenAI's ChatGPT

Copyright © 2024 Wildlife Tree, LLC. All rights reserved.

Designer: Allyson Randa

Photo Credits:

AdobeStock.com

Pixabay.com

Pexels.com

ISBN: 978-1-965081-00-6

This book meets **Common Core** and **Next Generation Science Standards.**